On Parity and Isospin

Greg Feild

April 11, 2017

Every step I take is with hesitation, and every new reflection makes me dread an error and absurdity in my reasoning.

-- David Hume

Abstract:

In this brief appendix to "The Sinister Universe", we demonstrate that there is one universal unit of charge, or one fundamental coupling constant responsible for all particle interactions.

The fundamental universal coupling constant is the charge to rest mass ratio of the electron; e/m_e.

We also demonstrate that the simultaneous measurement of the electron charge and mass is subject to the commutation relation

$$[q,m] = \hbar/i$$

Prolegomena:

A certain confusion, however, arose in science which cannot determine how far reason is to be trusted, and why only so far and no further, and this confusion can only be cleared up and all future relapses obviated by a formal determination, on principle, of the boundary of the use of our reason.

-- Immanuel Kant, 1783

philosophy is fun!

Preface:

Parity and isospin.

Boring and confusing!

… or two beautiful, bountiful constructions of human genius?

The answer, of course, is yes!

Who didn't skim over these bits in school? Parity is interesting, and a fairly straightforward concept as the 'mirror image' of an interaction; but tallying up the intrinsic parities of all the particles in an interaction for purposes of conservation is a tiresome and tedious chore. Plus, what does it really mean? How could a fundamental particle have such a quality? Now we think we know, of course, as this is the one of the subjects of this book!

My hat is off to the individuals who calculated and tabulated the intrinsic parity for all the known particles in the zoo. It seems a thankless task … but we thank them now! It takes a universe to build a model.

As for isospin, it just seems like something somebody (Werner Heisenberg) made up.

Since the proton and the neutron have quite similar masses, and "only" differ by this little thing called the electric charge, they must be kind of like the same particle!

Everything was made up at some point, of course, but this particular idea only leads to an approximate symmetry or conservation law, and thus it never seemed so compelling to me.

Now, I am a believer!

I have seen and grope toward the light.

The leptonic table:

LEPTONS   ANTI-LEPTONS

electron	electron neutrino	PARITY ⇔	electron antineutrino	positron
⇐	CHARGE	MASS ⇕	CHARGE	⇒
muon	muon neutrino	PARITY ⇔	muon antineutrino	anti-muon
⇐	CHARGE	MASS ⇕	CHARGE	⇒
tau	tau neutrino	PARITY ⇔	tau antineutrino	anti-tau
⇔	weak isospin	mass isospin ⇕	weak isospin	⇔

TABLE 1: The leptons and their interrelations; or the kleptogenesis of the leptoquarks.

Any lepton can be 'generated' from any other by the appropriate applications of the parity operator, the weak isospin operator, and our newly proposed 'mass isospin' operator.

Introduction:

In "The Sinister Universe" we demonstrated that matter particles and antiparticles have opposite parity because they *literally* spin in opposite directions. Leptons spin to the left and anti-leptons spin to the right!

We can now interpret particle-antiparticle annihilation as the cancellation of spins.

spin left + spin right == spin 0

We also demonstrated that the electron and the electron neutrino differ only by the electric charge e, and that the conservation of weak isospin currents for the basis (e,v) leads to the conservation of lepton number or the conservation of a vector quantity we called the electromagnetic charge, **L_em**;

L_em = e*hbar/2*c **s**   (1)

where **s** is the unit spin vector. (We then reached some half-baked conclusion about the conservation of baryon number ... a better theory is presented here.)

Conservation of baryon number is now believed to arise from the conservation of a similar 'generational' or 'mass' isospin current on the basis (e, mu, tau) leading to the conservation of the mass charge, L_m;

L_m = m*hbar/2*c   (2)

where m is the total relativistic mass energy of the particles involved in an interaction.

Of course, this is just the conservation of energy, electric charge, and spin wrapped up in a pretty new package with an additional or complementary 'explanation' attached!

These ideas are summarized in TABLES 1 and 2. The universal model of muon decay is shown in FIGURE 1.

Table of coupling constants, conserved currents, and charges:

Force	Coupling constant	Conserved current	Rotation basis	Conserved charge
Electricity	e/m_e	Weak isospin	e, v_e	e*hbar/2*c s
Gravity	m_e/e	Generational isospin	e, mu, tau	m*hbar/2*c

TABLE 2:. Note: m_e/e = m_v

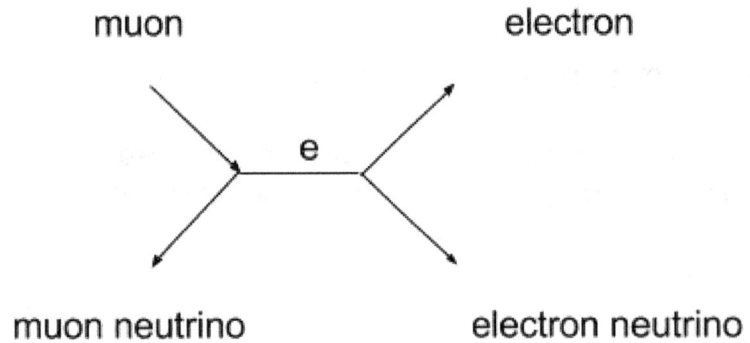

FIGURE 1: Muon decay. The propagator is most likely a generic virtual lepton, e/mu/tau.

From quarks to leptons:

We now know who ordered the muon ... the same individual who ordered the charm quark! In our model, they are one in the same.

The 'explanation' is that the muon (and tau) are necessary for the existence of heavy baryons. Of course, this may be begging the question ...

The muon and tau are also necessary for high energy particle interactions and decays.

We postulate that nature prefers decays where the kinetic and potential (or in our model, mass) energies of the initial and final state particles are approximately equal; or a kind of conservation of, or minimization of, 'the action'.

Wave your hands in the air.

The "four forces" of classical physics:

As shown in "The Sinister Universe", the classical relativistic Lorentz force is now

$$F = m*(e/m\_rest)*E + m*(e/m\_rest)*(v \times B) + m*F\_g + m*(v \times B\_g) \qquad (3)$$

where we are not going to worry about the relative plus and minus signs between the electromagnetic and gravitational terms for the moment. Factoring out the constant coupling charge, e/m_rest, we have

$$F = (e/m\_rest)(m*E + m*(v \times B)) + (m*F\_g + m*(v \times B\_g)) \qquad (4)$$

We see that the electromagnetic and gravitational forces now have exactly the same form and they both depend on and vary with the total relativistic mass energy of a particle.

For the electromagnetic terms, e/m_rest is *the* fundamental coupling *constant*.

The charge to rest mass ratio of a particle is now the fixed fundamental coupling constant, rather than e or alpha, and needn't be involved in calculations except as a multiplicative factor.

We note, when $m = m\_rest$, the electromagnetic and gravitational forces are totally 'decoupled'.

In general, the relative strengths of the four classical forces can now be written with respect to the electric force and the electric charge e, as follows; magnetism, e/c^2; gravity, e/m; and gravitational magnetism, e/mc^2.

Now, let's take a closer look at the new mass dependence of the Lorentz force as described by equation (4). For simplicity, we will focus only on the two electromagnetic terms. The force on a massive charged particle (e.g. an electron) due to a specified distribution of mass and charge is then

$$\mathbf{F} = (e/m_e)(m^*\mathbf{E} + m^*(\mathbf{v} \times \mathbf{B})) \tag{5}$$

If we consider the interaction between two identical electrons, we must assume one provides the field for the other, and equation (5) becomes

$$\mathbf{F} = (e/m_e)(m_1^*\mathbf{E} + m_1^*(\mathbf{v_1} \times \mathbf{B})) \tag{6}$$

Using the well known equations for **E** and **B** due to the second electron, we have

$$\mathbf{F} = (e/m_e)^2(1/R^2)(m_1 m_2(1/4\pi\varepsilon)\mathbf{r} + \mathbf{p1} \times \mathbf{p2} \times \mathbf{r}(\mu/4\pi)) \tag{7}$$

Some factoring yields

$$\mathbf{F} = (e/m_e)^2(c/R)^2(\mu/4\pi)(m_1 m_2 \mathbf{r} + (1/c^2)(\mathbf{p1} \times \mathbf{p2} \times \mathbf{r})) \tag{8}$$

For non-relativistic interactions, we can solve the relationship

$$E = p^2/2m \tag{9}$$

for the mass of our two electrons, finally yielding

$$F_{1,2} = (e/m_e)^2(c/R)^2(\mu/4\pi)(p_1^2 p_2^2/4E_1 E_2 + (1/c^2)p_1 p_2 \sin\theta) \tag{10}$$

assuming our two electrons are moving parallel to one another.

The purpose of this exercise was to demonstrate that the classical Lorentz force between two particles can be formulated solely in terms of their energy and momentum.

The classical coupling strength is now given by our new universal coupling parameter, $(e/m_e)^2$, which can be carried over unscathed and unmolested into quantum mechanical and QFT calculations!

Maxwell's equations:

We now turn to Maxwell's equations, the equations famous for uniting the forces of electricity and magnetism, and humbly suggest adding one final symmetry term; the magnetic current **J_m**, which is related to the electric current **J** by the simple relationship

$$\mathbf{J\_m} == (m/e)*\mathbf{J} \qquad (11)$$

Maxwell's equations, in vacuum, are now

$$\mathbf{Del} \times \mathbf{B} = \mathbf{J} + \partial \mathbf{E}/\partial t \qquad (12)$$

$$\mathbf{Del} \times \mathbf{E} = \mathbf{J\_m} - \partial \mathbf{B}/\partial t \qquad (13)$$

$$\mathbf{Del} \cdot \mathbf{E} = \text{rho} \qquad (14)$$

$$\mathbf{Del} \cdot \mathbf{B} = 0 \qquad (15)$$

We're not worthy! (And we might need a minus sign …)

As we have noted previously, the divergence of the magnetic field **B** is always zero since magnetic fields are generated solely by the relative or absolute motion of charge.

Quantum mechanical electromagnetic induction:

In the universal model, the electron neutrino is considered to be the fundamental unit of mass. It has a magnetic moment given by

$$\mu_\nu = e\hbar/2 m_\nu c \qquad (16)$$

The neutrino magnetic moment is assumed to arise from its spinning (point) mass in analogy with the usual arguments made for the magnetic moment of the electron as due to spinning charge.

In the sinister universe, the electron is thought to be an 'electrically excited' electron neutrino, and the mass of the electron can be written in terms of the electron neutrino mass and the value of the electric charge;

$$m_e = e \cdot m_\nu \qquad (17)$$

Similarly, the electron magnetic moment is

$$\mu_e = (1/e) \mu_\nu \qquad (18)$$

In our model, the electron magnetic moment is considered to be a result of spinning mass, just as in the case of the neutrino, and is *not* due to spinning electric charge.

The electric charge is created by the spinning electron mass through a process similar to that of classical electromagnetic induction!

Of course, the electron is stable because electromagnetic induction is quantized.

For the electron, Maxwell's equation (14) becomes

$$\boldsymbol{\nabla \cdot E} = e/m \qquad (19)$$

The uncertainty principle for mass and charge:

In this section, we will have to wave our hands a little more vigorously (!) than usual. Perhaps the arguments presented here will inspire someone to try and work it all out for real.

In "The Physical Principles of the Quantum Theory", Werner Heisenberg (he's back) derived the uncertainty principle for measuring the electric and magnetic fields in any arbitrarily small volume of spacetime;

$$\Delta E_x \Delta B_y \geq \hbar c/(\delta l)^4 \qquad (20)$$

In our electron model, the magnetic field of the electron is created by its spinning mass. This magnetic field then induces an electric field which is interpreted to be the 'point' electric charge.

The classical analogy is electromagnetic induction. Since the mass, spin, and charge of the electron are all time independent quantities, this implies electromagnetic induction is quantized in the particle realm.

So, we wave our hands and argue that we can make the following associations between the electric and magnetic fields of the electron and its electric charge and mass;

$$\Delta E \sim \Delta q$$

$$\Delta B \sim \Delta m$$

Then a miracle occurs!

And we see that one cannot measure the exact charge and mass of the electron at the same time

$$[q,m] = \hbar/i \qquad (21)$$

Group theory:

Remarkably (and necessarily, perhaps), the universal model retains the SU(3)SU(2)U(1) group structure of the standard model.

However, these groups do not generate 'exchange quanta'. In addition, the color charge is no longer the basis for the group SU(3). Instead, we propose the three charged leptons; the electron, the muon, and the tau.

The charged leptons are all identical except for mass, and thus seem appropriate as an isospin basis;

e = (1,0,0)

mu = (0,1,0)

tau = (0,0,1)

The relativistic mass and magnetic moment:

Our new electromagnetic charge, as shown in equation (1), is defined as the the product of the mass and the magnetic moment of a particle;

L_em = m*mu (22)

where

$m = m\_rest/(1 - v^2/c^2)^{½}$ (23)

and

$mu = (e*hbar/2*m\_rest*c)(1 - v^2/c^2)^{½}$ (24)

We can see that as the mass charge increases, the magnetic moment of a particle decreases, thus conserving and constraining the total electromagnetic coupling capability of all elementary leptons to be exactly the same, everywhere, and all the time!

Matchbook summary:

The universal Lagrangian is basically the Lagrangian of QED, except that we need to introduce the mass dependence of the charge using the energy operator

$$E = -i\hbar \partial/\partial t \qquad (25)$$

The gravitational and electromagnetic interaction terms in the Lagrangian would then look something like;

$$L\_interaction \sim (1 - e/m_{rest})(\psi^{bar} \gamma^{\mu} A_{\mu} \partial \psi/\partial t) \qquad (26)$$

Gauge theory:

We can now understand the origin of gauge invariance in QED.

The variable mass charge involved in an interaction is compensated for by a time dependent 'phase factor' in the wave function; $\exp(-iE*t)$.

physics is fun!

Conclusion:

Parity, isospin, conservation of momentum, mass, charge ...

Who comes up with this stuff?

Are parity and isospin real things? Is parity less real than electric charge or spin?

How did we even arrive at the notion that nature should be beautiful, symmetric, and governed by one single, simple underlying mechanism?

It takes a universe.

A universe of people collaborating across space and time!

The universal model:

It describes the universe; universally! But, we don't *have* to call it that ...

How about QED ++, or QFT +/-, or ... the Light Standard Model or Standard Model Lite or ...

a feild theory   :)

The unity of the physical universe:

Why is it then that, in spite of its evident disadvantages, the picture of the future [of physics] can hold its own against all the past?

It is simply the *unity* of the picture: unity of all separate parts of the picture, unity of space and time, unity of all experimenters, nations, and *kulturs*.

-- Max Planck,
"A Survey of Physical Theory"

Sunshine train:

    see the fields
    go rushin' by
    don't wonder
    where they've gone

    beauty is
    a thing of pleasure
    best be movin on ...

            -- Jon Butterworth

… and the physicist said, "O.K. then! Imagine your horse is a perfect sphere …"

Notes:

neon lights, a
Nobel prize, then

a mirror speaks,
the reflection lies

you don't have to
follow me

↺ ↺ ↺

the cult of personality

       -- Living Colour

www.ingramcontent.com/pod-product-compliance
Lightning Source LLC
Chambersburg PA
CBHW081135180526
45170CB00008B/3119